IMAGES
of America

CLEVELAND COUNTY AGRICULTURE

In this 1960s photograph, district supervisor Tom Cornwell points out with the Cleveland Soil and Water Conservation District's fair display how overgrazing decreases pasture productivity. The Cleveland Soil and Water Conservation District has always been supervised by a board of local farmers, and Cornwell grew cotton and raised Hereford cattle. The four other supervisors at the time were Yates Brooks, Cameron Ware, Harold Plaster, and Ralph Spangler. (Courtesy of Cleveland County USDA Service Center.)

ON THE COVER: In 1926, farmers gathered at the farm of M.S. Bean in Waco for a farm demonstration day on swine. County farm agent R.E. Lawrence stands to the far left with a tie and boater hat. (Courtesy of University of North Carolina Libraries.)

IMAGES
of America
CLEVELAND COUNTY
AGRICULTURE

Cleveland Soil and Water
Conservation District

ARCADIA
PUBLISHING

Copyright © 2016 by Cleveland Soil and Water Conservation District
ISBN 9781540200983

Published by Arcadia Publishing
Charleston, South Carolina

Library of Congress Control Number: 2015959967

For all general information, please contact Arcadia Publishing:
Telephone 843-853-2070
Fax 843-853-0044
E-mail sales@arcadiapublishing.com
For customer service and orders:
Toll-Free 1-888-313-2665

Visit us on the Internet at www.arcadiapublishing.com

To farmers of Cleveland County—past and present—we thank you.

CONTENTS

ACKNOWLEDGMENTS

Thank you, to the organizations, families, and individuals who provided photographs, insight, and expertise for this project. Previous staff of the Cleveland Soil and Water Conservation District (SWCD) and USDA Soil Conservation Service took photographs and documented farmers' attempts to practice good land stewardship. The discovery of a box of their old photographs started this endeavor. A special thank-you is extended to the previous staff and the current Cleveland SWCD Board of Supervisors, each of whom was supportive from day one: Randy McDaniel, Ted Wortman, Michael Underwood, Roger Eaker, and Sherri Greene. A special thank-you goes out to Myron Edwards and Randy McDaniel for providing background information related to photographs, Sherri Greene for editing and reviewing the text, and Krista Parker for scanning and organizing photographs.

Thank you, to my wife, Natalie Bishop, who provided enthusiasm, interest, and encouragement, and to my parents, Randy and Connie Bishop, who have always supported my writing and have proofread many a text, including this one.

All of the author's proceeds from the purchase of this book go to the Cleveland County Farmland Preservation Program, a program that increases awareness of agriculture, which is still the county's leading industry. To the best of my knowledge, all photographs in this book were taken in Cleveland County or feature Cleveland County residents.

—Stephen Bishop
Cleveland SWCD

INTRODUCTION

Before the spread of photography in the early 1900s, the land of present-day Cleveland County had long been cultivated—for hundreds of years by settlers and thousands of years by Native Americans. In fact, the Broad River was once the Line River. It formed the rough boundary line, if not skirmishing ground, between two rival and powerful tribes, the Catawba and Cherokee. Both tribes relied on hunting, gathering, and cultivating crops. The bottomlands around the Broad River provided plenty of fertile soil.

Since we have no photographs, try picturing Cleveland County at this time, just 400 to 500 years ago. Black bears, red wolves, and eastern cougars would hunt not only deer but also buffalo and elk (think Buffalo Creek). Grasslands bustled with bobwhites and turkeys. Native Americans often burned these grasslands to prevent trees and shrubs from growing. Forests would have been full of chestnut trees, now long gone due to blight. The trees themselves would have cackled with the sound of bright green birds called Carolina parakeets, now extinct. The streams would have been clearer, cooler, and full of brook trout. Overall, the landscape would have looked quite wild and different.

One familiar sight in that landscape might have been corn. Maize was a major crop for the Catawba and Cherokee. Most southeastern tribes planted several varieties of flint and flour corn. Early explorers commented on the abundant granaries and corncribs, chock-full of dried corn on the cob. In 1775, explorer and trader James Adair described an old Catawba cornfield over seven miles long. The Catawba had already been decimated by smallpox, but Adair states that they once must have been strong and numerous to clear and work a field that large.

Native Americans planted a variety of other crops, including beans, squash, sunflowers, and tobacco. They often interplanted crops within the same field, and southeastern tribes became well-known for the "three sisters" method of planting corn, beans, and squash together. Men cleared fields by girdling trees and burning brush. Women planted, weeded, and harvested crops. Some southeastern tribes had community fields in which each family tended and worked a plot. Other tribes had scattered family fields, and each family sent a portion of its harvest to the community storehouse or granary. Food saved in community storage was used for ceremonial feasts and times of hardship.

Europeans settlers arrived in earnest in the 1700s. By many accounts, the Catawba were quite friendly. Unfortunately, this only hastened their decline, as smallpox spread quickly. In 1682, the Catawba totaled 4,800 members. By 1784, their population had plummeted to 250. The Cherokee had a more hostile reputation. Benjamin Cleveland, the namesake of Cleveland County, led several campaigns against the Cherokee. By 1830, the Cherokee were forced to midwestern reservations along the route known as the Trail of Tears.

Early trade among Native Americans and settlers often focused on agriculture. Although southeastern tribes hunted buffalo and turkey, keeping domesticated livestock was a new concept. Horses, cows, sheep, goats, and chickens were widely traded for. Many European settlers traded for and grew Native American crops. Both corn and tobacco became important crops for the earliest European settlers in Cleveland County. Corn provided corn meal and mash for livestock—and the liquor still.

Many settlers to Cleveland County had already lived in Virginia or Pennsylvania. Some worked as indentured servants for seven years to pay off their passage to North America. Others owned farms but relocated as Virginia and Pennsylvania became more populated. Often, they traveled the Great Wagon Road to Charlotte and then moved into the North Carolina backcountry. With no roads into the backcountry, these settlers had to travel by trail. And unlike the Coastal Plain, where gently winding streams and rivers allowed some navigation, the western Piedmont and Foothills of North Carolina afforded no such luxury, with streams full of drops and waterfalls. Travel for settlers in Cleveland County remained difficult for many years. In fact, the first dirt road was not built across Cleveland County until 1852.

Settlers were often Scotch-Irish or German. Historical accounts describe Scotch-Irish settlers as great frontiersmen, fiercely independent and unafraid of skirmishing. Scotch-Irish were a large portion of the Overmountain Men who defeated the British at Kings Mountain. German settlers were generally considered more peaceful, community-oriented, and better farmers. Both groups had certain farming tendencies. For instance, Germans used more oxen than mules or horses. Scotch-Irish kept more sheep. Germans typically took more time clearing fields, felling trees with axes. In contrast, Scotch-Irish girdled trees and planted around standing stumps. Despite differences in farming methods, the shared legacy of Scotch-Irish and German settlers is still easily seen. Look no further than the prevalent names of Cleveland County. Scotch-Irish names include Blanton, Lattimore, McMurry, Kendrick, Hendrick, McDaniel, Alexander, Yarbro, Roberts, and Thompson. German names include Beam, Dellinger, Plonk, Lutz, Lineberger, Wortman, Weber, and Weaver.

European settlers also brought slaves. According to the 1860 census, the total number of slaves in Cleveland County was 2,131, or around 15 percent of the population. Nearly half of these slaves worked for the Froenberger paper mill on Buffalo Creek. Unlike in the eastern part of the state, where slaves worked on large plantations, slaves often worked in different roles in the backcountry. In *A Journey in the Back Country*, Fredrick Law Olmsted states, "Of the people who get their living entirely by agriculture, few own [slaves]; the slave holders being chiefly professional men, shop-keepers, and men in office, who are also land owners, and give divided attention to farming." Local historian J.R. Davis wrote that a few small farmers could afford to keep around 5 to 10 slaves, but many had no slaves. Large-scale plantations were unfeasible with a lack of nearby markets for farm products. Still, the crops that slaves brought along had a lasting impact on the county. Sweet potatoes, cane sorghum, and cotton would eventually become staples on Cleveland County farms. Cotton would become the county's triumph.

Before the Civil War, cotton was seldom grown in Cleveland County. Most farmers grew what they needed to survive, and the idea of farming a cash crop was still many decades away. For fiber, farmers grew flax and kept sheep. Animal hides were also used for clothing. Corn was the most widely grown product in the county. All of the corn was used on the farm to feed livestock and for corn meal. Wheat and oats were important crops. Farmers cultivated more fertile areas, and grazed cows or sheep on poorer areas. Since Cleveland County was formed from parts of Lincoln and Rutherford Counties in 1841, the 1850 *Census of Agriculture* provides a good first glimpse of what was grown and raised in the county. In 1850, the county contained 145 oxen; 343 mules; 2,188 horses; 2,415 milk cows; 4,689 beef cattle; 7,820 sheep; and 16,311 hogs. Farmers produced 14,035 pounds of wool; 91,775 pounds of butter; 9,336 pounds of beeswax and honey; 335,572 bushels of corn; 64,682 bushels of oats; 36,952 bushels of wheat; and 43,909 bushels of sweet potatoes. Only 321 bales of cotton were produced.

The Civil War affected agriculture in several ways. Agriculture was extremely labor intensive. Many farmworkers were either killed, wounded, or freed and never returned to fieldwork. With laborers gone during the war, some chores went undone despite the best efforts of those who remained home. Soldiers who did return found farms in various states of neglect. In 1914, based on conversations with local residents who lived through the Civil War, historian J.R. Davis wrote,

But the greatest problem that confronted the people of Cleveland in 1865 was restoring the economic life of the county. When the farmers returned home from the war they found their fences torn down and their houses dilapidated. Their land was becoming dotted with gullies and their labor was demoralized. The fifty-seven water wheels which were converting the water of the streams into power in 1860 were standing idle and many of the dams burst.

In a letter dated June 28, 1868, Emeline Putnam from Patterson Springs writes to her sister and brother-in-law, describing some of the hardships: "The crops are bad in this part of the country. No harvest at all. Hardly worth cutting. The corn crops are very bad. Some folks just got done planting. . . . It is the sickest summer we have had for some time." Two years later, corn production was still down nearly 40 percent from 1860 levels.

After the war, turbulence and unrest also descended on the county. Conflicts and power struggles over political, social, and economic issues spilled into agriculture. In 1868 and 1870, some farmers in the southern part of the county refused to rent land to African Americans who voted Republican, the party of Lincoln. Even in 1908, forty-three years after the war, the *Charlotte Observer* reported a Shelby cotton gin received a letter from "nightriders." The letter demanded that the gin stop buying cotton until prices rose above 12¢ a pound, or else the gin would "go up in smoke."

To say the least, Cleveland County, like the entire South, was a war-torn country. Many years passed before the county rebounded, but eventually it began to thrive. In the early 1870s, three small cotton mills opened with secondhand equipment along the First Broad River. By 1880, Cleveland County produced 6,126 bales of cotton, a tenfold increase from 1870. One of those first three mills, Cleveland Mills, expanded and relocated to a state-of-the-art facility in Lawndale in 1888. With new local demand, cotton acreage continued to increase.

Cleveland County also became a leader in cotton because production shifted westward from the worn-out soils of the east. Depleted soils and erosion were also problems in Cleveland County. Good farmers manured and rotated fields; some experimented with horizontal planting and hillside ditches, the forerunning practices to contour planting and terraces. Still, it was standard practice to turn over the land in the fall and allow it to sit uncovered till spring harrowing and planting. Researchers believe the county lost on average four to seven inches of topsoil during the late 1800s and early 1900s (to build back an inch of topsoil takes 400 years). In many fields, the brown sandy topsoil was completely washed away, exposing the red clay subsoil. These areas, known locally as "red lands," were often synonymous with cotton production.

The good news is that red clay, although more difficult to work and cultivate, was relatively fertile. Cotton production continued to increase as farmers began using commercial fertilizers, improved cotton varieties, and soil conservation practices. Although cotton was becoming a cash crop, farms remained diverse and self-reliant—with gardens, orchards, hogs, poultry, and cows to supply the table. In a 1903 article, the *Cleveland Star* states,

> About half the farmers of the county raise cane. . . . Those who use the new evaporators make a remarkably good grade of syrup. It will be safe to say that every man in the county who owns or rents land raises a garden full of fine cabbage, and in the fall many of them make one or two barrels of kraut. Apples, Peaches, Pears, Quinces, Alums, Cherries, Strawberries and Blackberries grow in great abundance.

In 1909, there were 884 acres grown in cane sorghum. Another 1,291 acres were in cowpeas. The cowpeas were used for hay, picked green and shelled for the table, or harvested dry and consumed over winter.

In the early 1900s, farming remained the primary occupation in Cleveland County. Nearly 40 percent of Cleveland County's population farmed (that is down to less than two percent today). The county developed a reputation for forward-thinking and progressive farmers. S.H. Colwick

9

formed a farmers union, the first of its kind in the state, in Boiling Springs in 1905. The union's purpose was to share farming ideas and techniques and to improve educational opportunities and economic conditions for local farmers. By 1906, Colwick organized 11 other local groups to form the Cleveland County Farmers Union. This led to the creation of a state farmers union in 1907. Among their accomplishments, these unions successfully lobbied lawmakers for the creation of farm life schools and a rural farm credit system.

Farms at this time were often small by today's standards. The 1910 *Soil Survey of Cleveland County* states that the average-sized farm was around 65 acres "with numerous farms of 20 to 30 acres." Already, outright farm ownership was beginning to decline, with farmers working more and more rented land. Much of the work on these small farms was done by the family, and only larger farms hired help. More competition for labor from cotton mills meant many small farms could not afford to hire workers.

Cotton was boom or bust. Many farmers in the early 1900s wrestled with how much land to plant in cotton, as cotton prices were notoriously volatile. A 1926 article from the *Spartanburg Herald* states,

> With the present situation prevailing, cotton selling low, everybody offering advice and none confident to act upon it, farm leaders—those who think in front of the calendar—have reached the conclusion that in the upcoming year that chickens and cows will supplant cotton as the cash crop in Cleveland County. Which doesn't necessarily mean there will be no cotton planted next year, but every available acre will not be given over to cotton.

The Great Depression made problems even worse. In 1934, Congress passed the Bankhead Cotton Control Act, a quota program to reduce the amount of cotton on the market and keep prices from dropping. Some Cleveland County farmers were skeptical at first, but by the 1940s, J.S. Wilkins, the county extension agent, stated that Cleveland County farmers are "almost 100 percent for controlled cotton production."

And for good reason. In the 1940s, Cleveland County was the leading cotton producer in the state. Times were good. War World II was over, and the profit from cotton meant farmers could afford more luxuries: automobiles, indoor plumbing, electricity, tractors, hay balers, and pull-type combines. Labor-saving devices meant a single farmer could farm more land, and many farmers encouraged their children to seek other opportunities.

Still, as farmers know, disaster can strike in an instant: droughts, floods, price drops, disease, and pests. In 1949, the boll weevil surged into Cleveland County. In one year, the county lost over $5 million of cotton (or $50 million adjusted for inflation today) to the insect. According to an article in the *Cleveland Star*, collections were taken up by schoolteachers because so many schoolchildren were going hungry. Despite efforts to control the boll weevil with new pesticides, some cotton gins were closing down and moving out of the county by the mid-1950s. More and more people would leave farming for more stable jobs.

Since then, Cleveland County agriculture has seen other rises and falls. In 1964, the county was dotted with 213 grade-A dairies, but only five remain today. Currently, the poultry industry is booming, with more poultry houses being built each year. Although farming is not as prominent as it once was, Cleveland County is still by far the biggest agricultural county in the Foothills, with poultry, cattle, and field crops leading the way. As of the latest *USDA Census of Agriculture*, farmers in Cleveland County generate over $120 million in sales each year.

Increasingly, people want to reconnect with farms and farmers. Perhaps some folks want to reconnect with nature and the land, and farmers, of course, should always be stewards of the land. For others, farms represent idyllic places—places where patience, faith, and hard work prevail. Hopefully, these photographs will provide a glimpse at the triumphs and trials of farm life in Cleveland County many decades ago. A lot has changed since these photographs were taken. What has not changed is that farming remains hard, and often unheralded, work.

One

BARNS, BUILDINGS, AND FARMHOUSES

Today, people can drive to stores and buy whatever is needed for day-to-day life. However, for many past Cleveland County residents, going into Shelby to shop was once an event reserved for special items the farm could not produce. In the early 1900s, the journey into town meant harnessing horses, hitching up buggies and wagons, and traversing crude roads. It was not a quick trip to the store and back.

Early-1900s farms had to be self-reliant. At any given farm, an array of outbuildings was dedicated to various farm enterprises and activities. Farms had a mixture of corncribs, blacksmith shops, chicken coops, tater mounds, well houses, outhouses, smokehouses, pigpens, springhouses, lumber sheds, silos, and barns—small, big, and grandiose. At the center of this village of barns and outbuildings was the farmhouse. Sometimes, the farmhouse had to double as an outbuilding too. For instance, one photograph in this chapter shows an impressive brick farmhouse from 1938, once owned by the Falls family on Carpenters Grove Church Road near Belwood. As grand as that house is, notice that raw cotton is being stored under the carport. Likewise, many farmers stored raw cotton on the front porch temporarily before hauling it to the gin. Early farms made do with what they had, and a porch was as good a place as any to keep cotton dry.

Eventually, farms became more specialized, and the diverse mixture of outbuildings was replaced with bigger facilities specializing in one or two farm enterprises. Equipment sheds became as common as general-purpose barns. The photographs in this chapter not only span this transition but also reflect differences in class. Most early farmhouses were small and modest, not likely to be photographed. That is always something to remember when looking at early farm photographs. Photographs were usually reserved for occasions or places out of the ordinary.

Two farmhouses sit at the foot of King's Pinnacle, in the vicinity of present-day Lake Montonia Road. This 1920s photograph depicts the rural landscape of the past, with plenty of open space between farmhouses on an old dirt road. (Courtesy of Pauline Mauney Kellam.)

The family of Lawson and Penola Kendrick pose in front of their white clapboard farmhouse, a typical middle-class farmhouse of the time. A cotton farmer, Lawson built this house in 1897, and this photograph was taken around 1910. Later, 10 family members lived together in this three-bedroom house. The house still stands on Pleasant Hill Church Road, near Patterson Springs. (Courtesy of Natalie Edwards Bishop.)

In the late 1890s, a traveling photographer passed through northern Cleveland County, photographing families in front of their farmhouses. The above image features the large family of David Decatur in front of their house on Kistler Road. Decatur grew cotton, corn, and wheat. The photograph below shows the home of Henry and Clarence Warlick on Oak Grove–Clover Hill Church Road. Notice the hitching post in front of the woman and also the side door behind the boy on the bicycle. That door led to the parson's nook, where traveling preachers came in and spent the night without disturbing the family. (Both, courtesy of Myron Edwards.)

This photograph was taken around 1980 but shows an old farmhouse built in the late 1700s. Crowder Mauney built the house, located on Oak Grove–Clover Hill Church Road, and later sold it to Amos Edwards. The house was actually three stories, with a large basement story. A cook's house was once located to side of the house. In those days, most houses had detached kitchens, or cook's houses, to reduce fire hazards in the main house. (Courtesy of Myron Edwards.)

In this 1940s photograph, a woman showcases a milk cow with a simple farmhouse in the background. Notice that neither the house or barn is painted. After the Great Depression, county farm agents promoted painting farmhouses and outbuildings not only to prevent decay but also to increase morale. (Courtesy of Cleveland County Cooperative Extension.)

Three children and a man pose in chest-high cotton. This c. 1950 photograph most likely depicts a tenant house. The tenant family, including these children, tended and picked the cotton in return for board and a share of the cotton profit. Tenant farming was a common practice, with wealthier landowners providing the land and capital and tenant farmers providing the labor. (Courtesy of Cleveland County Cooperative Extension.)

15

The Holland House in Boiling Springs is a classic example of a traditional farmhouse. The T-shaped design has a two-story front, with two central chimneys, and a one-story back wing housing the kitchen. This house became the headquarters for the Gardner-Webb College farm in the 1950s and still stands today on College Farm Road. (Courtesy of Gardner-Webb University Archives.)

The Falls family once farmed hundreds of acres of cotton in upper Cleveland County. In 1938, cotton was stored under the carport of their brick farmhouse before being hauled to the gin. (Courtesy of Cleveland County Cooperative Extension.)

Here are two classic white farmhouses. The above photograph, from 1950, shows the Holland House in Boiling Springs towering in the background. The farmhouse below belonged to Tom Cornwell, one of the first district supervisors for the Broad River Soil Conservation District (which later became the Cleveland SWCD). Now gone, this house once stood on North Lafayette Street, a little north of the roundabout. The photograph was taken in 1938. (Above, courtesy of Gardner-Webb University Archives; below, courtesy of North Carolina State University Libraries.)

These two photographs portray a mainstay on Cleveland County farms, the general-purpose livestock barn. In the 1940s photograph above, a little girl hides behind a calf in the barnyard. In the background is a common A-frame barn, the standard barn type for the era. These barns had stalls along a central hallway and a hayloft above. Before the hay baler, farmers pitched loose hay or bundled hay from a binder into the loft. In the photograph below, Will McCurry and his family stand in front of a newly built barn. Notice the loose hay already in the loft. (Above, courtesy of Cleveland County Cooperative Extension; below, courtesy of Robert Jones.)

Until tractors became common, draft animals were an important part of any farm. A good barn to maintain these animals was important. During periods of heavy fieldwork, mules were kept in stalls overnight to quickly harness in the morning. The farmer in this c. 1940 photograph is proud of two young mules. (Courtesy of Cleveland County Cooperative Extension.)

In the background of this c. 1950 photograph are two crib-type outbuildings. Cribs were often raised on fieldstone foundations to store livestock feed, usually dried cob corn. To keep feed from molding, cribs had plenty of chinks, gaps, and openings for air circulation. (Courtesy of Gardner-Webb University Archives.)

In 1950, a paint crew gives this barn a face-lift. Paint crews often traveled from community to community offering farmers their services. Red was the most popular paint color, but some farmers were more thrifty and used old motor oil to paint their barns. Painting helped prolong the lifespan of a barn. Sometimes farmers earned a small sum from a paint job, allowing companies like Coca-Cola or Pepsi to use their barns as billboards. (Courtesy of Gardner-Webb University Archives.)

This work crew is setting poles for a pole barn. As farms began using more mechanized equipment and increasing herd size, large pole barns were more cost-efficient than traditional A-frame barns to build. They provided plenty of storage space for hay and equipment at a much lower cost per square foot to build. This photograph was taken in 1950, showcasing the construction of a new pole barn at the college farm at Gardner-Webb College. (Courtesy of Gardner-Webb University Archives.)

Two men perform their version of a high-wire act. They moved from pole to pole atop planks with absolutely no safety equipment. This pole barn was part of Gardner-Webb College's farm, which provided vocational training in innovative farming techniques and beef, pork, and poultry for the college cafeteria. (Courtesy of Gardner-Webb University Archives.)

Here stands the finished product. In 1950, this pole barn dwarfed the general-purpose barns found on most farms. In fact, the barns are so large that the draft horses at the left edge of the photograph look small. (Courtesy of Gardner-Webb University Archives.)

Two

KING COTTON

In the early to mid-1900s, Cleveland County was a leader in cotton production. Over 40 cotton gins once operated here, with a gin or two in every community. Ginned cotton was baled and shipped to warehouses or directly to textile mills. By 1910, eleven textile mills operated in the county to buy cotton, and any excess was shipped to mills in Gastonia or Charlotte. The Dover Mill opened in 1923 and used 3,000 bales each year. Several cottonseed-oil mills took leftover seed. With the nearby demand for cotton, farmers dedicated more and more acreage to the crop, with nearly 70,000 acres planted in cotton in 1925. One seed company, Coker Pedigree Seed Company out of Hartsville, South Carolina, leased a gin from Morgan and Company in Shelby to gin cotton for seed production. Eventually, its variety of Cleveland Big Boll became one of the most popular cotton varieties on the market.

Several factors contributed to cotton's downfall. In the early 1900s, the boll weevil arrived. Like most insects, boll weevil populations surged or slumped in certain years due to favorable or unfavorable weather. Farmers could control weevil damage in normal years with cotton dust, usually calcium arsenate or DDT. A 1923 *Cleveland Star* headline proclaimed, "Shortage of Arsenic Puts Cotton Farmers at Mercy of Weevil." In some years, even with pesticides, populations were impossible to control. The insect did incredible damage in 1923 and 1949. By the mid-1950s, the boll weevil was developing resistance to many insecticides. In addition, cotton prices fluctuated greatly and farmers' ability to produce a bumper crop sometimes worked against them, with overproduction driving prices lower. Government quota programs were enacted to stabilize prices with varying degrees of success and controversy. Also, competition for labor from textile mills made finding good workers to hoe and pick cotton difficult. Finally, modern fertilizers and global markets allowed cotton production to shift westward to once-marginal cropland in Texas and California and eventually overseas.

As of 2015, only two Cleveland County cotton gins remain in operation: Boggs' Gin in Fallston and Hamrick's Gin above Boiling Springs.

In the 1920s, more and more land was planted in cotton. Clearing land was difficult, laborious work, mostly done with axes and draft animals. To clear this land in Lattimore, the crew had the help of an early crawler tractor. According to Cleveland County farm agent R.E. Lawrence, in 1921, the government distributed over one million pounds of dynamite to farmers at cost to help clear stumps. (Courtesy of Warren Crowder.)

This 1920s photograph shows the same land from the previous photograph, now completely planted in cotton. Notice that several stumps remain in the field. Many farmers planted around standing stumps. Also, early terraces were often steeper and left unplanted, like those photographed here. Later farmers began using broad-based terraces that could be planted and cultivated as well. (Courtesy of Warren Crowder.)

In this 1930s photograph, Anthony Beam stands in a field of cotton, bolls open and ready to pick. In 1924, Beam started the D.A. Beam Company, a slaughter and rendering facility in Shelby. One of its advertisements reads, "We pick up dead or cripple cows, horses, mules, & hogs from your farm free of charge." The company also sold seed and Super-Gro fertilizer. (Courtesy of Irene Camp.)

In the 1940s, cotton was king. For decades, Cleveland and Robeson County battled back and forth for the title of most bales produced. In November 1928, the *Robesonian* newspaper proclaimed, "It has come at last. The foot-of-the-mountains county of Cleveland had ginned more cotton by November 14th, than its only competitor Robeson." By the late 1940s, Cleveland County had pulled away as the undisputed cotton champion, producing 64,846 bales in 1947. In second place, Robeson County produced 40,622 bales. Unfortunately, the boll weevil hit hard in 1949, and cotton production was never the same. (Courtesy of Earl Scruggs Center.)

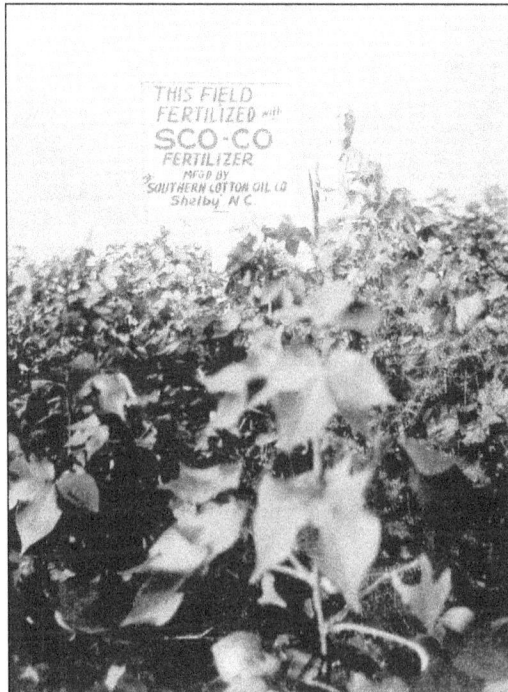

Walking in high cotton was once a source of pride for farmers—and fertilizer representatives. Above, W.A. Crowder stands in a cotton field in the mid-1920s. Crowder lived and farmed in Lattimore and sold fertilizers for AA Fertilizer Company. According to the *Cleveland Star* in 1927, Crowder set a record for cotton production, producing 235 cotton bales on 200 acres. The article notes that Crowder used nine mules, "giving 26 acres of cotton per mule." At left, a salesman advertises for SCO-Co fertilizers. The fertilizer was manufactured by the Southern Cotton Oil Company in Shelby. (Above, courtesy of Warren Crowder; left, courtesy of Ron Wilson.)

When the boll weevil arrived in the 1920s, farmers began dusting cotton to attempt to control it. In 1923, two hundred farmers met at the courthouse in Shelby to talk about boll weevil control and decided to buy a train carload of cotton dust. The most commonly used dust was calcium arsenate, which farmers spread by hand in the 1920s. Some farmers spread early in the morning, when the dew was still on plants, to help the dust stick better. Other farmers added a little molasses to the mixture. This 1926 photograph shows a demonstration on dusting. Cleveland County farm agent R.E. Lawrence stands in the center of the photograph with his hands on his hips. (Photograph by Cotten Cutter, courtesy of University of North Carolina Archives.)

Here, a Cleveland County farmer uses a mule-drawn duster to dust cotton. This was much faster and more effective than dusting by hand, and a farmer could dust about 10 acres in a day. By far the fastest way to dust cotton was by airplane, which became more common on larger cotton farms. (Courtesy of Cleveland County Cooperative Extension.)

Picking cotton was a family affair. Many small farms could not afford to hire outside labor, so the whole family picked, including children. Labor was the largest cost associated with picking cotton, so a big family was an asset during harvesttime. In the 1942 photograph above, the family must be starting, because a lot of cotton is left to pick—and they are still smiling. Below, this photograph shows a close-up of a cotton boll. There were two ways to pick cotton: to pick the cotton clean out of the boll or to pull the whole boll and let the gin clean it. (Above, courtesy of Earl Scruggs Center; below, courtesy of Cleveland County Cooperative Extension.)

Few people have fond memories of picking cotton. It was rough on the back and hands. In this 1920s photograph, notice these pickers in a Cleveland County cotton field remain bent over. Good pickers picked around 300 pounds a day. According to the *Cleveland County Soil Survey*, in 1910, Cleveland County cotton pickers were paid an average of 50¢ per 100 pounds of cotton picked. (Photograph by Cotten Cutter, courtesy of University of North Carolina Archives.)

In this 1920s photograph, Sam Weathers and his family stand in a cotton field in the Poplar Springs community. They produced twelve 500-pound bales on eight acres using International Rainbow fertilizer. (Courtesy of Robert Jones.)

This young boy's sack has some cotton inside. Into the mid-1900s, schools in Cleveland County let out during the fall for several weeks so children could help harvest cotton. If it rained enough to prohibit field work, schools opened up again. Some children could pick around 100 pounds a day. (Courtesy of Larry Wilson.)

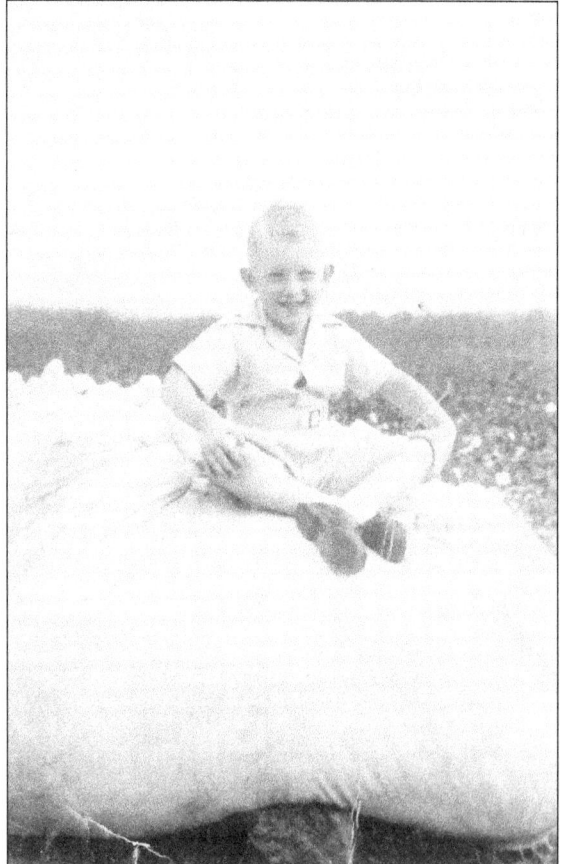

In this early-1950s photograph, Keith Warlick sits atop a loaded cotton sheet. Pickers dumped cotton onto sheets, which were bundled and tied up at the end of the day. Most farmers had a set of cotton scales so that they could weigh the cotton sheets before loading up the wagon or truck and heading to the gin. (Courtesy of Myron Edwards.)

A group of farmers observes a single-row, pull-type cotton picker. Mechanical cotton pickers arrived on the market much later than combines. The performance of early mechanical cotton pickers like these was not always up to farmers' standards, as they left a lot of cotton in the field. Sometimes farmers sent workers into the field to pick by hand the cotton left behind. This cotton picker is a Dearborn, an implement line associated with Ford tractors. Hardin-Dixon Tractor Company, on Washington Street in Shelby, sold Dearborn implements. (Courtesy of Earl Scruggs Center.)

The arrival of larger self-propelled cotton pickers eliminated much of the labor on cotton farms. And new herbicides meant farmers did not need laborers to weed cotton either. Now, one farmer can tend hundreds, if not thousands, of acres on his own. As with most field crops, growing cotton has become a capital-intensive, not labor-intensive, enterprise. (Both, courtesy of Earl Scruggs Center.)

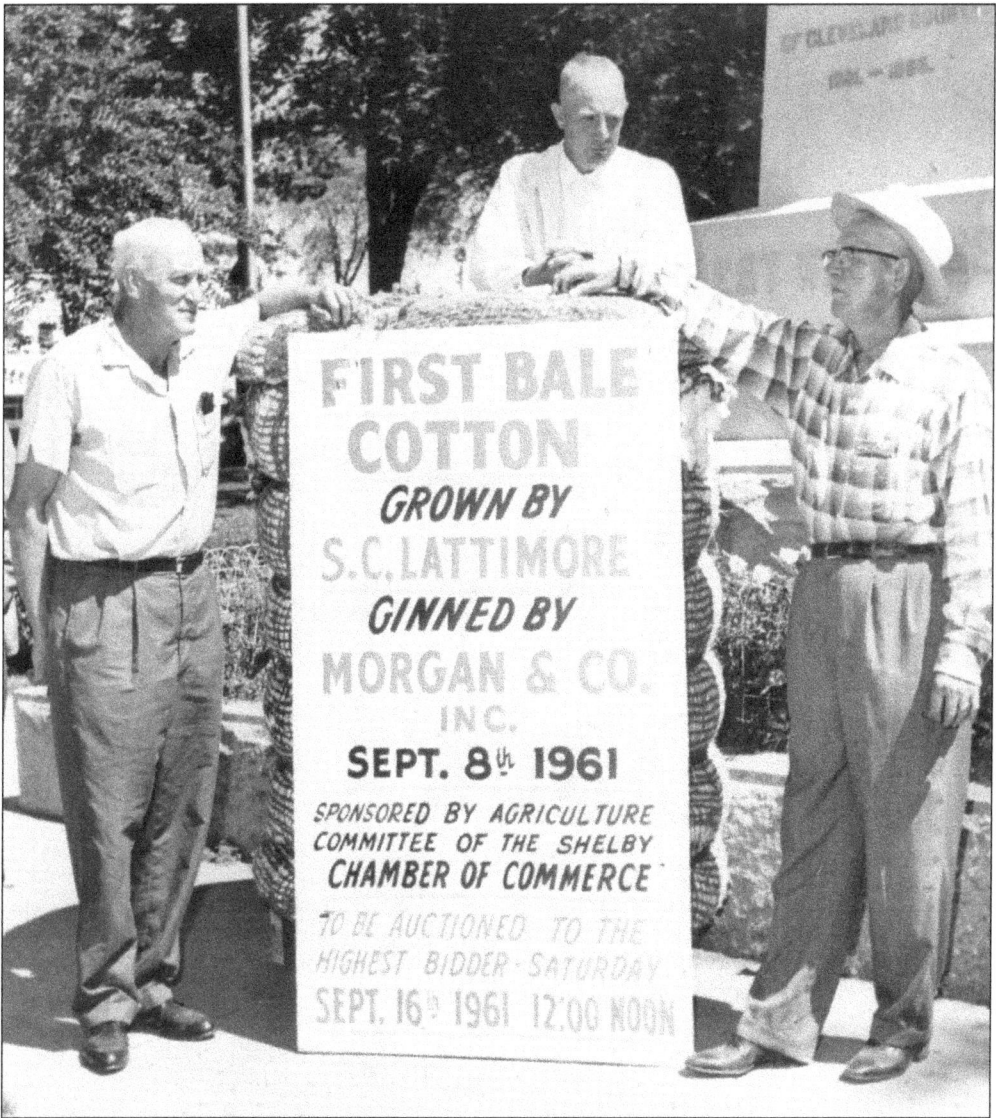

In 1961, the "first bale of cotton" ceremony was held on the town square in front of the county courthouse in Shelby. Morgan and Company ginned cotton and sold farm supplies near the intersection of Highways 180 and 150 in Shelby. Later, the Cleveland County Fair began purchasing the first bale at twice the market value to display each year. (Courtesy of Cleveland County Cooperative Extension.)

The family of Sam Conley poses for a photograph after winning the award for the first bale of cotton in Cleveland County in 1929. The award came with a cash prize and was very competitive. In 1929, Cleveland County led the state in bales of cotton produced. (Photograph by Ellis Studios, courtesy of Earl Scruggs Center.)

In 1930, T.F. Sellers from Kings Mountain won the award for the first bale in Cleveland County. The description on the back of the photograph reads, "T. F. Sellers of King's Mountain, North Carolina, winner of the $200 as first prize in Cleveland contest by producing 5,084 pounds of lint cotton on five acres of land, 1930." (Photograph by Robert Worth Shoffner, courtesy of North Carolina State University Libraries.)

A line of wagons waits to off-load at the P.D. Herdon Gin in Kings Mountain on October 25, 1934. During harvesttime, long lines at the gin were common. The sign beneath the unloading shed reads "Farm relief—only on this side—Thursday." In 1933, in the midst of the Great Depression, President Roosevelt signed the Agricultural Adjustment Act and the Farm Relief Bill. As part of these programs, farmers limited acreage of certain crops, like cotton, in return for a government payment. The purpose of the program was to prevent overproduction and stabilize cotton prices. (Photograph by E.C. Blair, courtesy of North Carolina State University Libraries.)

During cotton harvest, gins were the busiest places in town. Here, the F.S. and J.J. Crowder Gin in Lattimore has ginned bales lined up, ready for shipping to a mill or warehouse, and a line of trucks waiting to off-load. This gin was built in 1948; it closed in 1979. The ginning equipment was bought and relocated to Texas. (Courtesy of Warren Crowder.)

After a long wait, a family helps unload cotton at a Cleveland County cotton gin—possibly the Polkville Gin or Boggs' Gin in Fallston. The 1950s Ford truck is loaded down. (Courtesy of Larry Wilson.)

Trucks are lined up ready to off-load at the F.S. and J.J. Crowder Gin in Lattimore. During harvest season, many gins remained open all night, working through the line of waiting trucks. Many a farmer slept overnight in the truck cab at the gin. Once the last truck was unloaded for the day or night, ginners only had a few hours to sleep before starting the process again the next day. (Courtesy of Warren Crowder.)

This photograph from the late 1930s displays the cotton drier at McKinney's Gin in Cleveland County. USDA engineers began creating designs for cotton driers in the 1920s to aid the ginning process. By 1932, the Gullet Gin Company had begun manufacturing this Government Vertical Cotton Drier. (Courtesy of North Carolina State University Libraries.)

These two photographs were taken in 1939, and both display the same Cleveland County cotton gin. According the backs of the photographs, the name of the gin was the New Deal Gin, most likely because it received stimulus funding from the New Deal legislation during the Great Depression. (Both, courtesy of North Carolina State University Libraries.)

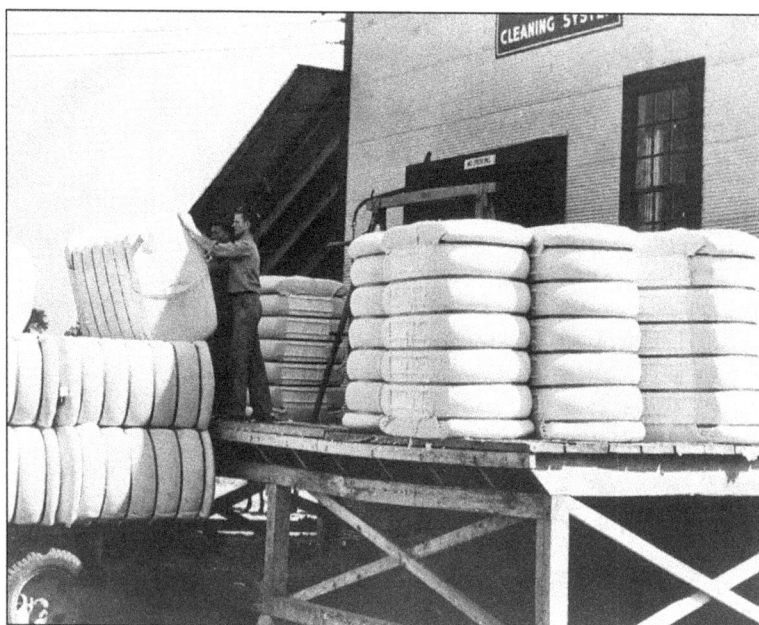

Two workers load a truck with cotton bales for shipment to a mill or warehouse. Moving these bales was hard work, as each weighed around 500 pounds. (Courtesy of Earl Scruggs Center.)

In this 1920s photograph, a potential buyer inspects and grades a sample of cotton. Cotton was graded on color, fiber length, and debris. Sometimes, the boll weevils left stains on the cotton and reduced the quality. Longer-staple cotton brought a higher price. (Photograph by Cotten Cutter, courtesy of University of North Carolina Libraries.)

Although many cotton bales went directly to textile mills, others were shipped to warehouses for temporary storage. This 1920s photograph shows workers off-loading ginned cotton bales at a warehouse in Shelby. Notice the man with a tie is grading a sample of cotton to determine the quality. (Photograph by Cotten Cutter, courtesy of University of North Carolina Libraries.)

A $10 million cotton crop is something to celebrate. The banner on this truckload of cotton bales reads "10,000,000.00 Crop Grown in Cleveland County." In the era of the boll weevil, this 1950s accomplishment was truly something to brag about. (Courtesy of Earl Scruggs Center.)

This 1950s photograph shows a more modern warehouse receiving a shipment of freshly ginned cotton. The Dover family owned this bonded warehouse in Shelby to store cotton. The warehouse was located on Marion Street behind the present-day Ora's Supermarket. (Courtesy of Earl Scruggs Center.)

In the early to mid-1900s, the relationship between agriculture and industry in Cleveland County was mutually beneficial. Textile mills needed cotton, so farmers grew cotton. One of the first textile mills in the county was Cleveland Mills, built in 1888 in Lawndale. The mill operated its own cotton gin, buying cotton directly from farmers in the early 1900s. These two photographs were taken inside Cleveland Mills, with the view in the above photograph overlooking the weaver room. (Both, courtesy of Earl Scruggs Center.)

Cleveland Mills closed in 2001. It is difficult to imagine the amount of cotton that traveled through and was transformed by the maze of spindles that once operated in the textile mills of Cleveland County. (Courtesy of Earl Scruggs Center.)

Three

APPLES TO ALFALFA

The label "cotton farmer" was misleading. In the early to mid-1900s, cotton was a farm's cash crop, but it was by no means the only crop. Farms were diverse. Most farmers grew a mixture of grains and field crops and raised several types of livestock. For example, B.M. Jones owned a 130-acre farm in Lattimore. According to his 1940 Farm Conservation Plan, he had 8 acres of corn, 28 acres of cotton, 8 acres of cowpeas, 8 acres of wheat, 5 acres of oats, 5 acres of barley, 18 acres of lespedeza, 18 acres of crimson clover, 14.3 acres of kudzu, 8 acres of pasture, 32.6 acres of woods, and 3.1 acres of miscellaneous crops. He kept 4 mules, 4 dairy cows, 30 chickens, and 3 hogs. Although he grew nine different crops and kept a variety of livestock, B.M. Jones was known as a "cotton farmer."

The other crops grown on Jones's farm were likely for home or farm use. The grains would have been used for feed. Some wheat and corn would have been milled for flour and cornmeal. According to the conservation plan, the cowpeas, crimson clover, and lespedeza were plowed under as soil builders. The kudzu was mowed for hay, which was common practice then.

The photographs in this chapter provide an example of the diversity of farming undertaken in Cleveland County. Cotton garnered the most attention, but other crops were also important. In 1929, a *Cleveland Star* article stated, "They [county farmers] are great on corn and cotton, but big as these items are, they are but incidental. The county's gold mine is in the hen house, the cow barn, the vetch and alfalfa field, and the sweet-potato curing house." Sweet potatoes, apples, peaches, cane sorghum, wheat, oats, barley, and cabbage were common "minor" crops found throughout the county.

Nearly every farm had a home orchard, but commercial orchards were also common. Many sent apples to the White House Vinegar plant in Lincolnton. Notice the thick stand of soil-building crimson clover between the rows of R.L. Jolly's apple trees in 1966. Crimson clover was an annual clover, and this variety reseeded itself each year. This stand of clover is 12 years old. (Photograph by Samuel Jenkins, courtesy of Cleveland County USDA Service Center.)

In this 1964 photograph, Everett Lutz (left) enjoys an apple with D.S. Weaver, the state director of the cooperative extension. Lutz owned an apple orchard in Belwood and was president of the North Carolina Apple Growers Association. (Courtesy of North Carolina State University Libraries.)

Humid conditions make fruit growing a challenge in the Southeast. In this photograph, a crew uses a pull-type spraying rig to protect peach trees from disease and insect pressure. (Courtesy of Cleveland County Cooperative Extension.)

This photograph depicts Wayne Ware's peach orchard in Kings Mountain in 1938. Notice Kings Pinnacle in the background. This photograph was taken near the present-day Rhodesdale Farm Stand. Between the young peach trees, the ground has been cultivated, which was once common practice in orchards to reduce weed competition. (Courtesy of North Carolina State University Libraries.)

By the early 1900s, most farms were using commercial fertilizers in addition to animal manures to maintain fertility. Mixing fertilizers allowed farmers to consolidate multiple fertilizer applications into one, reducing the amount of fieldwork needed to apply fertilizers. Farmers often mixed rock phosphate and nitrate of soda. This 1925 photograph shows a fertilizer mixer on display at the Cleveland County Fair. (Courtesy of Cleveland County Cooperative Extension.)

George Edwards (right) receives a shipment of fertilizer in Lawndale. Nitrogen was often the limiting factor for plant growth in the early 1900s, and soda nitrate was the most commonly applied nitrogen source. Higher-grade ammonia fertilizers replaced soda nitrate in the later part of the century. (Courtesy of Myron Edwards.)

48

County farm agent R.E. Lawrence examines some sweet potatoes with a Cleveland County farmer. During the 1920s, when this photograph was taken, sweet potatoes were big business. Nearly every community had a curing house, or "tater barn," for cooperative storage and marketing of sweet potatoes. In 1922, the Kings Mountain Sweet Potato Cooperative consisted of 40 farmers. (Courtesy of University of North Carolina Libraries.)

At first sight, this might look like a field of rye, with the grain almost overtopping the children. However, this field of wheat belonged to Forest Crowder (man in hat) of Lattimore. The photograph was used as an advertisement for Virginia Carolina Fertilizer Company. Today, wheat is bred to grow much shorter to prevent lodging and reduce clogs in the combine. (Courtesy of Warren Crowder.)

R.L. Plonk stands in a field of rye in 1966. Rye, wheat, barley, and oats were all planted on Cleveland County farms. Rye was used on farms for livestock feed. It performed better in poorer soils than other small grains. (Photograph by Samuel Jenkins, courtesy of Cleveland County USDA Service Center.)

This photograph was taken on May 7, 1951. K.O. McKinney of Grover points out to Joe Craver how strip-cropping helps manage the rain. May is the typical planting month for cotton, so the bare strips in this field were most likely planted in cotton soon after this photograph was taken. (Photograph by Samuel Jenkins, courtesy of Cleveland County USDA Service Center.)

Strip-cropping was a common practice in Cleveland County in the mid-1900s. Instead of planting a whole field in one crop, farmers alternated strips of grain crops with row crops like cotton or corn. The grain crops held the soil better, slowing erosion. In 1951, district conservationist Joe Craver examines a field at Charles Dixon's farm with Dixon's wife alongside. (Photograph by Samuel Jenkins, courtesy of Cleveland County USDA Service Center.)

Taken in May 1964, this photograph shows W.C. Bingham looking over his farm. The strip of bare ground was planted in a row crop for the summer. The oats were just beginning to head out and were likely harvested in early June. (Photograph by Samuel Jenkins, courtesy of Cleveland County USDA Service Center.)

Strip-cropping was common for several decades in Cleveland County. It was not until the adoption of no-till farming in the 1970s that strip-cropping began to decline. This photograph was taken in 1962. Cotton, corn, and some type of summer sod are all being grown at the same time in the same field. (Photograph by Samuel Jenkins, courtesy of Cleveland County USDA Service Center.)

A Fordson tractor pulls a binder. Binders had a ground-driven sickle-bar mower that cut the grain and a mechanism that tied and bundled the grain before dropping the bundles throughout the field. The bundles were then picked up, collected, and hauled to a thresher. (Courtesy of Earl Scruggs Center.)

Roy Cochran harvests a field of barley in 1972 with his Minneapolis Moline combine. The combine earned its name by combining the function of the thresher and binder into one machine. The first combines were pull-type combines that cut around six feet at a time. Modern combines cut 40 to 50 feet at a time. (Photograph by Jim Boggs, courtesy of Cleveland County USDA Service Center.)

Before combines, threshing was a time-intensive and laborious step in the process of harvesting grains. The first thresher in Cleveland County was a "groundhog" thresher in the Washburn community in 1849. It was hand-cranked, with two men cranking on each side of a conclave that beat the grain out of the straw. The belt-driven thresher in this 1927 photograph is an upgrade over the hand-crank days. The two men to the left are sitting and standing on the front of the thresher, most likely a Wood Brothers Individual Thresher or Geiser thresher. Both brands were sold together with Fordson tractors. For instance, Chas Eskridge in Shelby sold Fordson tractors and Wood Brothers threshers. By the 1940s, pull-type combines replaced threshers. (Courtesy of Warren Crowder.)

Legume crops were grown on Cleveland County farms for forage, soil-building, and feed. It is hard to believe now, but in 1923, sixteen vetch and alfalfa clubs existed in the county, totaling over 400 farmers. The *Cleveland Star* states, "Every farmer has his vetch and oats patch and is making enough feed for his horses and cattle without pulling fodder." Vetch, cowpeas, and lespedeza were often plowed under as a "green manure" to add nitrogen to the soil. Above, three men examine a stand of vetch that will soon be plowed under. Below, a farmer stands in a thick field of alfalfa. (Both, courtesy of Cleveland County Cooperative Extension.)

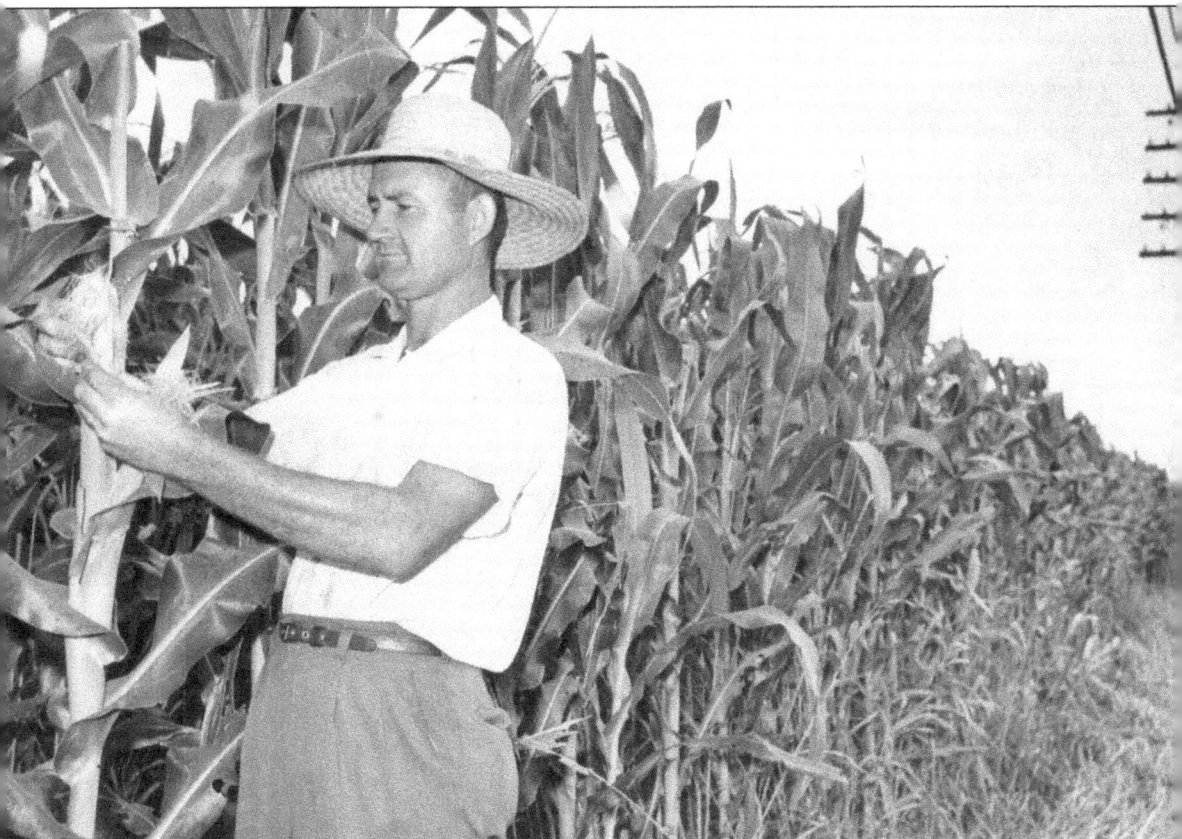

Corn was the second-most common field crop grown in the county. Nearly all of it was destined for livestock, either through harvested grain or silage. Here, a Cleveland County farmer examines an ear of corn. (Courtesy of Cleveland County Cooperative Extension.)

In 1961, the corn in A.W. Greene's field near Rehobeth Church is nearly twice his height. The Greene Dairy grew all the feed needed for the herd on the farm: 210 acres of corn, 100 acres of small grains, and around 100 acres of pasture and hayland. (Courtesy of Greene family.)

Susan Weathers (far right), Cleveland County's first home demonstration agent, organized tomato clubs in various communities. Tomato clubs taught girls over the age of 10 how to tend a garden and can. Each girl planted a tenth-acre garden and sold the extra produce, canning what did not sell. (Courtesy of Cleveland County Cooperative Extension.)

Penola Kendrick holds a large watermelon. Her husband, Lawson Irvin Kendrick, farmed cotton, but the family always kept a large garden and several hives of bees to help pollinate the garden. Appropriately, *penola* is actually the Chickasaw word for cotton. (Courtesy of Natalie Edwards Bishop.)

The homestead hog was a mainstay on farms as well. The most common breeds of hogs were the Berkshire and Poland-China. In the 1920s, the Shelby Poland-China Breeders Association and the Waco Berkshire Association helped promote these two breeds. Both are black and are good foragers. These photographs show how pigs were often left to roam the woods and take advantage of the acorn crop during the fall. (Both, courtesy of Cleveland County Cooperative Extension.)

Other farmers kept pigs in pens. Above, Zeb Holland works on his pigpen. Below, a group of high school students watches a sow with her piglets. (Above, courtesy of Jim Holland; below, courtesy of Cleveland County Cooperative Extension.)

This herd of swine is hogging down what appear to be sweet potato vines. Farmers often hogged down the patch of sweet potatoes used to produce slips as well as the main sweet potato fields after harvest. Hogging down sweet potato fields was promoted as a way to reduce the sweet potato weevil. This photograph was taken in 1926 at the M.S. Bean farm in Waco. (Courtesy of University of North Carolina Libraries.)

Four

BUTTER AND MILK

After the boll weevil disaster in 1949, the government encouraged farmers to diversify. Many farmers turned to dairying, which had already been an important part of farm income for decades. In the early 1900s, cream was the most valuable dairy product. In 1910, North Carolina had six creameries, two of which were located in Cleveland County: the Mooresboro Creamery and the Shelby Creamery. Both made butter, which was then shipped across the state and the Eastern Seaboard. Shelby Creamery's butter won first prize at the state fair in 1922, a year in which it paid local farmers nearly $200,000 for cream (equivalent to $3 million today). Each creamery had pickup routes throughout the county to collect cream. Farmers kept the skimmed milk for home use or fed it to livestock.

Milk later became a valuable product, with farmers receiving higher prices for higher grades of milk. Most small farms sold grade-C or -D milk for manufacturing or cooking purposes only. In 1953, there were 150 grade-A dairies in Cleveland County and 600 selling lower-grade milk. In 1940, after building an evaporating plant in Statesville, Carnation located a receiving station in Shelby for farmers to deliver milk in 10-gallon cans. Just down the road on Grover Street, the Carolina Dairy also bought milk from local farmers. Farmers around Kings Mountain often sold milk to Sunrise Dairy in Gastonia. Larger dairies began selling milk in bulk to distributors like Sealtest in Charlotte.

As with cotton, the price of milk was volatile. "Milk wars" were common, as large and small distributors continuously undercut one another. In response, politicians appointed a milk commission to examine the problem. According to the *Gaston Gazette*, a milk war was occurring in Shelby in 1972, when Ab Wolfe of Sunrise Dairy in Gastonia said of a proposed regulation, "It discriminates against the little distributor. The big boys are going to eat us up. That's all there is to it." By the late 1970s, regulators had set minimum prices paid to farmers for raw milk and prohibited distributors from providing incentives to retailers. A decade later, the government began paying farmers to quit dairying altogether.

This photograph shows the Carnation receiving station on Lee Street in Shelby. The station opened on June 20, 1940, to such high anticipation that 340 farmers supplied 13,000 pounds of milk on the first day. From Shelby, the milk was shipped to the Statesville evaporating plant to be processed into condensed milk. (Courtesy of Cleveland County Cooperative Extension.)

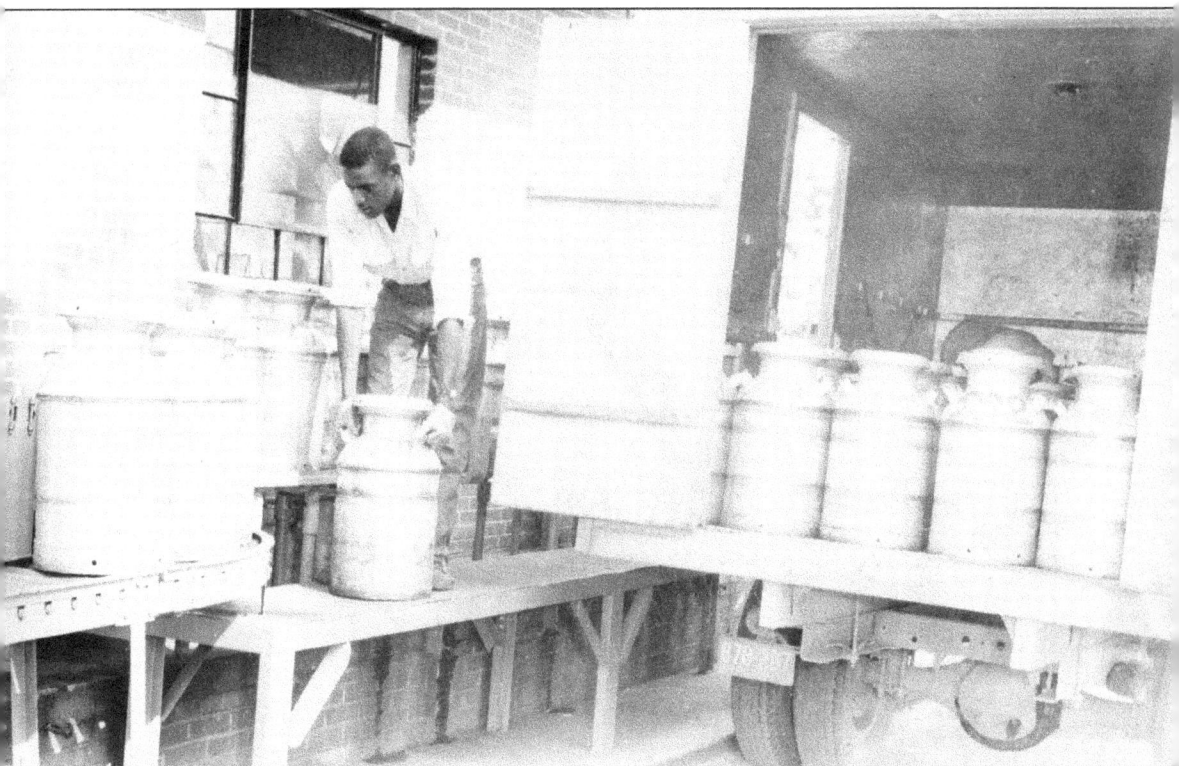

A man unloads 10-gallon cans full of raw milk. For many years, this is how milk was delivered to pasteurizing and distribution plants. Each distributor had service routes along which it picked up milk from local farms. Farmers would receive a check every two weeks for the milk they supplied. (Courtesy of Cleveland County Cooperative Extension.)

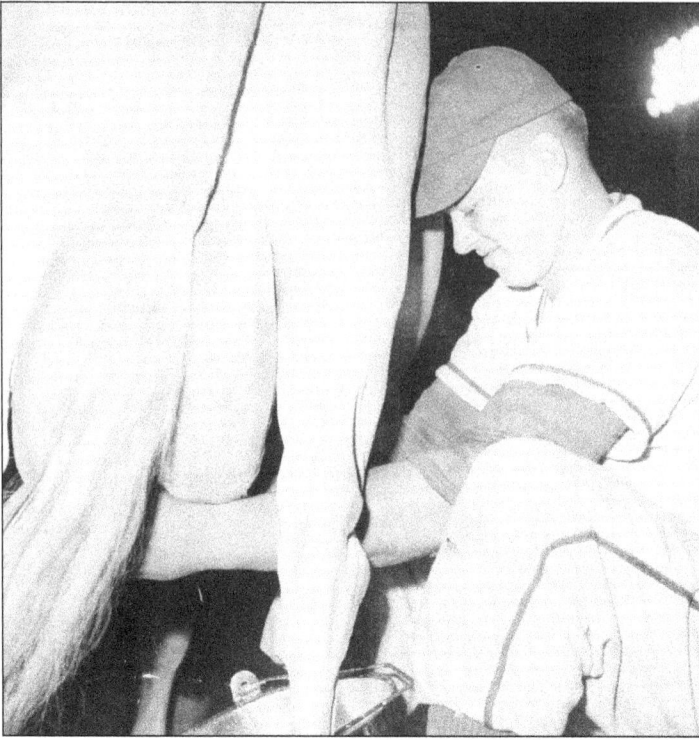

An unidentified baseball player tries his hand at milking. In the late 1940s, some farms began installing electric milkers to speed up the process. County farm agents at the time were promoting parlor upgrades not only to increase efficiency but also to enable farms to sell grade-A milk. As herds became larger, electric milkers became the norm. (Courtesy of Cleveland Cooperative Extension.)

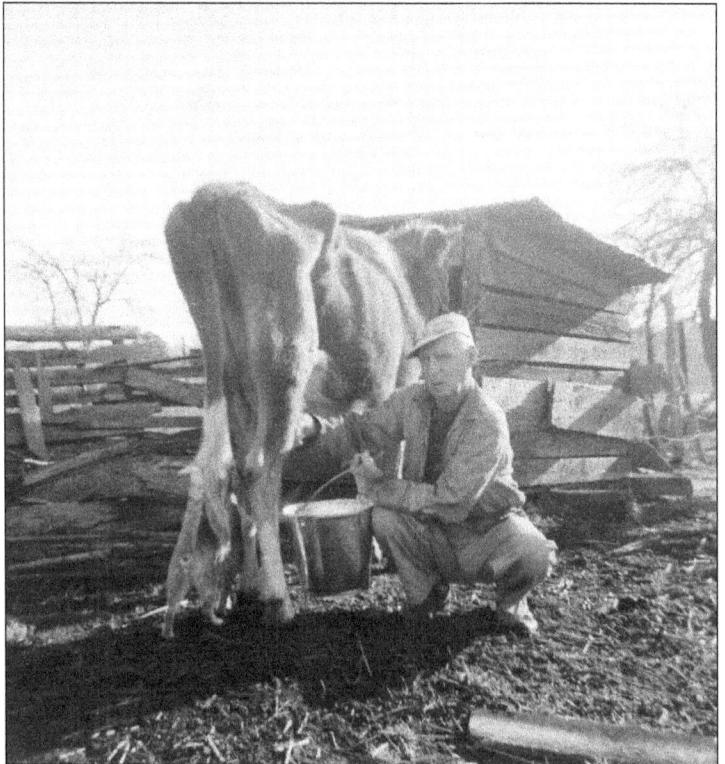

Zeb Holland milks his cow. Even if farms did not sell milk, many kept a cow to milk for home use. Throughout his life, Holland continued to keep a homestead cow for truly fresh milk. (Courtesy of Jim Holland.)

White clothes were the dress attire for dairy showing, and this boy is showing what appears to be a Brown Swiss. Compared to Guernseys, Jerseys, and Holsteins, the Brown Swiss was one of the minor dairy breeds found in Cleveland County. According to the *Cleveland Star*, other minor dairy breeds in the county were the Ayrshire, Dutch Belted, Devon, and Red Polled. This photograph was taken in 1937 featuring the Cleveland County Dairy Demonstration Team. (Courtesy of North Carolina State University Library.)

Harold Randall stands beside a Guernsey cow in 1942. Guernseys look similar to Jerseys but have white spots, a flatter face, and a fuller frame as distinguishing features. The Randall family near Kings Mountain was well-known for their herd of Guernsey cows. (Courtesy of North Carolina State University Libraries.)

Above, a line of Holsteins eats at the Greene Dairy in 1961. Below, young ladies pose with a Holstein calf at the Cleveland County Fair. The Holstein breed, with black and white patterns, is now iconic. In Cleveland County, however, Guernseys and Jerseys were once more common than Holsteins, which slowly started gaining favor in the 1950s and 1960s with breeding improvements and, reputedly, a better temperament than the other breeds. (Above, courtesy of Greene family; below, courtesy of Cleveland County Cooperative Extension.)

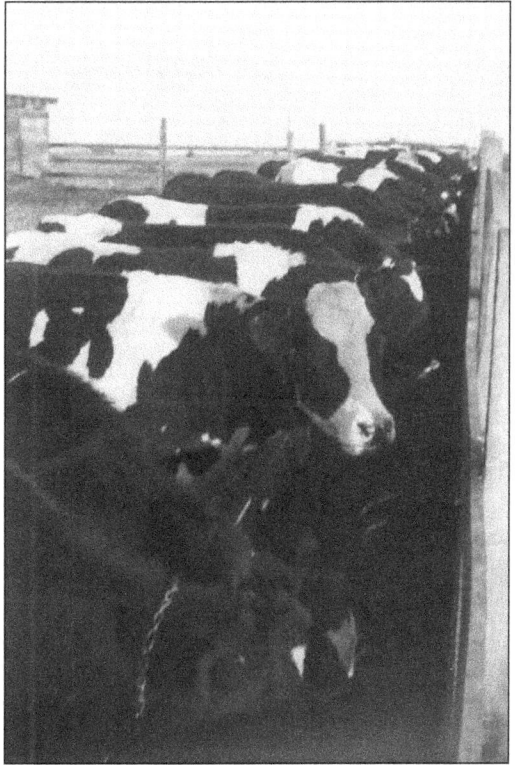

In these two photographs from 1959, a line of Holsteins chows down on feed at the Greene Brothers Dairy. In 1953, A.W. Greene had seven Guernseys, 10 Holsteins, and four Jerseys. He decided to reduce his cotton acreage and expand his dairy operation. By 1965, the farm had a herd of 198 Holsteins, with Greene's sons Albert, Philip, and Randall helping run the operation. According to the *Shelby Star*, Albert and Randall milked at 5:00 a.m. and 5:00 p.m. every day, which took two hours for each milking, while Philip handled the feeding. (Both, courtesy of Greene family.)

In this 1940s photograph, a young boy stands beside a prizewinning Jersey cow. The Jersey is one of the smaller breeds of dairy cows. Despite their smaller size, Jerseys produce more milk per pound of body weight than many larger breeds. (Courtesy of Cleveland County Cooperative Extension.)

In this 1940s photograph, a judge is examining a line of cows. He has a lot to choose from at this junior dairy show. Dairy shows were very competitive, as farmers used the shows to highlight their stock. A champion bull or cow could generate a great deal of income for breeding purposes. (Courtesy of Cleveland County Cooperative Extension.)

This photograph is from a dairy show at the Gardner-Webb College farm, and the handlers are most likely college students. The college farm started in 1950 to provide vocational training in farming and food for the college cafeteria. (Courtesy of Gardner-Webb University Archives.)

Two shirtless boys are not quite in dairy show attire. Still, their Jersey cows are something to be proud of. (Courtesy of Gardner-Webb University Archives.)

These two photographs are from around 1950. Above, a student (right) and his brother (both barefooted) stand with two Guernsey cows. Below, three Jersey cows are looking toward a classroom building in the background. (Both, courtesy of Gardner-Webb University Archives.)

Two young cowhands are seen in these photographs. Above, a boy poses with a Jersey calf. Below, a barefooted child stands beside a calf that is four times his size. It is hard to imagine a parent today letting a child stand barefooted beside a calf. (Above, courtesy of Gardner-Webb University Archives; below, courtesy of Cleveland County Cooperative Extension.)

A dairy show occurs at the college farm at Gardner-Webb College. Although O. Max Gardner, the namesake of the college, was best known as a politician, serving as governor from 1929 to 1933, he was also a leading farmer in the county and a true advocate for agriculture. He was often invited to speak, not on political matters, but on the intricacies of cotton farming and innovative farming methods to groups throughout the state. During the 1950s, the university honored his commitment to agriculture by creating the college farm. (Both, courtesy of Gardner-Webb University Archives.)

Two typical herds of dairy cows are depicted in these photographs. In the 1940s, the average herd was around 10 to 20 cows. Most farms at the time were very diverse, milk being one of many income sources. Dairy herds were also an important source of manure to spread on fields as fertilizer. (Above, courtesy of Cleveland County Cooperative Extension; below, courtesy of Cleveland County USDA Service Center.)

Two Gardner-Webb College students hug a polled Jersey cow. In polled cows, horn production was bred out. Many farmers, however, noticed a drop-off in milk production from horned to polled cows, and horned cows remained standard on dairy farms. (Courtesy of Gardner-Webb University Archives.)

Dairy cows have a reputation for being high-strung. They were near the top of the list of ways to be injured on a farm. Farmers dehorned cows to protect against injury, not only to themselves but also to other cows. (Courtesy of Earl Scruggs Center.)

This aerial photograph, taken around 1950, shows Pinnacle View Farm on Bethlehem Road near Kings Mountain. The Randall family owned this farm and was well-known for their herd of Guernsey cows. The farm had two silos, a milking parlor, a barn, grain bins, and equipment sheds. (Courtesy of Earl Scruggs Center.)

The model in this fair booth portrays an early-1950s dairy. CleCo Farm had 22 Guernsey cows, 10 heifers, and 2 bulls. It produced "Golden Gurnsey" milk, which was reputedly higher in vitamins than milk from other breeds. (Courtesy of North Carolina State University Libraries.)

These boys participated in the Cleveland County calf chain. Hopefully, this heifer calf grew up and gave birth to another heifer calf to continue the chain. The silos in the background were once very common on Cleveland County dairy farms; silage provided a higher-quality feed than hay. (Courtesy of Cleveland County Cooperative Extension.)

Corn and sorghum were used for silage. In these two photographs, a self-propelled Fox silage chopper cuts the sorghum into small pieces and blows the silage into a truck following alongside. Once full, the truck took the silage to a silo for storage. (Both, courtesy of Cleveland County USDA Service Center.)

Here are two examples of bunker silos on Cleveland County dairy farms. After packing silage into a bunker, farmers covered the bunker, preventing air from getting in. Now, most dairy farms rely on bunker or pit silos because they are easy to build and pack with modern front-end loaders and heavy machinery. (Both, courtesy of Cleveland County Cooperative Extension.)

Five

HAY, HEREFORDS, AND ANGUS

Today, fescue grass and Angus cattle are typical sights on Cleveland County cattle farms. Prior to 1950, however, neither were common. The ideal pasture at the time was a mix of Ladino clover and orchard grass. In 1947, Tom Cornwell of Shelby grew the first five acres of fescue in the county and used a pull-type combine to harvest the grass seed. Cornwell was a member of the North Carolina Foundation Seed Producers. Members grew new varieties of crops and forages to increase seed stock for distribution to farmers. According to the *Charlotte Observer*, after Cornwell's efforts to increase seed stock, some fescue was being grown in most parts of Cleveland County by 1949.

Cornwell also grew cotton and raised Hereford cattle. Herefords were the most common breed of beef cattle in the county in the early to mid-1900s, with Shorthorns and Angus lagging behind. Herefords received a premium price at cattle sales, a fact of which farmers were certainly aware. In the 1950s, two livestock sales took place weekly in Shelby, one on Mondays at the Dedmon's Sale Barn and the other on Wednesdays at Shelby Sale Barn. Eventually, at cattle sales, the tide began to shift in favor of Angus. Herefords had been intentionally bred for smaller animals that fattened early, as many of the photographs in this chapter illustrate. Although desirable at the time, this trait eventually led to the Hereford's downfall from the dominant breed in Cleveland County and across the country. The meatpacking industry began demanding leaner and trimmer animals, and the Angus breed fit that bill.

Today, Cleveland County ranks 10th in the state in head of cattle, most of which are some variation of Black Angus. Although finding a purebred Hereford cow is not as easy as it once was, finding a Black Baldy, a black cow with a white face, is easier. The Black Baldy is a cross between a Hereford bull and an Angus cow.

Cattle in the mid-1900s were much shorter and stockier than today, as these photographs of Polled Herefords illustrate. Above, a Gardner-Webb College student poses with his Polled Hereford bull before a cattle show at the college farm. Below, a line of Polled Hereford steers awaits judging at a cattle show in Shelby. Cattle trends later changed, with taller and larger cattle preferred. (Above, courtesy of Gardner-Webb University Archives; below, courtesy of Cleveland County Cooperative Extension.)

John Hendrick watches his herd of Herefords in 1955. Hendrick kept cattle and was also one of the first turkey farmers in the county. (Both, courtesy of North Carolina State University Libraries.)

The boys in these c. 1940 photographs provide a good measuring stick for the size of the Hereford breed. Even crouched down, the two boys to the left are almost taller than steers. (Both, courtesy of Cleveland County Cooperative Extension.)

The Cleveland County Fair once had a massive barn, pictured above, for livestock shows. Below, around 1940, a man poses with an Angus bull after a cattle show at the Cleveland County Fair. (Above, courtesy of Earl Scruggs Center; below, courtesy of Cleveland County Cooperative Extension.)

Located on present-day John Crawford Road, the farm of Knox and Shaw Surratt was once an excellent example of pasture management. As this photograph from 1972 indicates, cattle were rotated between pastures to prevent overgrazing, even if it meant moving cattle across a road. This line of cattle, mostly Charolais, is calmly moving through the intersection of two dirt roads. (Photograph by Jim Boggs, courtesy of Cleveland County USDA Service Center.)

In the 1950s, the Soil Conservation Service began working with farmers to build farm ponds. Many ponds in Cleveland County were built through the Soil Conservation Service, either to impound a gully to stop erosion or to provide a stable water supply for a farm. Knox and Shaw Surratt used a farm pond as a central part of their pasture management system. (Both, photograph by Jim Boggs, courtesy of Cleveland County USDA Service Center.)

A mixed herd of Black Angus and Herefords graze on fescue pasture at the farm of Earle Hamrick, three miles north of Shelby. (Photograph by Elmer Turnage, courtesy of Cleveland County USDA Service Center.)

Carl Francisca of Kings Mountain showcases a bull named Afterglow on July 11, 1949. This bull would be considered small by today's standards. (Courtesy of North Carolina State University Libraries.)

By the 1960s, Kentucky 31 fescue was the dominant pasture and hay grass in Cleveland County. A cool-season grass, fescue grows well in the fall and spring but goes dormant during summer. Many farmers experimented with warm-season grasses. This is the second cutting of hay from a field of Pensacola Bahia grass at Roy Dedmon's farm in 1962. For warm-season production, Bahia grass later fell out of favor for Bermuda grass. (Photograph by Sam Jenkins, courtesy of USDA Service Center.)

A farmer cuts hay around 1960 in a Cleveland County field. (Courtesy of Cleveland County USDA Service Center.)

A farmer and his wife are combining grass seed using a pull-type combine around 1955. This combine has a sacking platform instead of a grain bin. The woman, sitting on the platform, tied the sacks and slid them down the shoot. The sacks fell into the field and were collected later. (Both, photograph by Joe Craver, courtesy of Cleveland County USDA Service Center.)

A member of the North Carolina Foundation Seed Producers, Tom Cornwell grew the first Kentucky 31 fescue in the county. He was also a supervisor for the Cleveland Soil Conservation District. The district distributed the seed to farmers for better grazing productivity. Above, Cornwell stands in a field of fescue ready for harvest on June 6, 1951. Below, Cornwell is harvesting seed from the first fescue field in Cleveland County in 1946. (Both, courtesy of Cleveland County USDA Service Center.)

Six

THE CHICKEN OR THE EGG

Currently, broiler houses generate nearly $50 million in gross income each year in Cleveland County, continuing a long tradition of poultry production in the county.

In the 1920s, the county farm agent, R.E. Lawrence, was a major advocate for poultry. At the time, the boll weevil was just beginning to move into North Carolina. Lawrence pushed farmers to increase poultry flocks as an alternate source of income in case the boll weevil did considerable damage. Lawrence's plan was for each community to form a poultry cooperative. According to the *Cleveland Star*, to form a cooperative, at least 10 farmers in each community had to pledge to keep "not less than 100 hens, each of the same breed, and build an up-to-date poultry house 12 ft x 24 ft to house the 100 hens." Lawrence's efforts were successful, with cooperatives started in the Waco, Bethlehem, Casar, New House, Cedar Grove, and Broad River communities. In 1923, eggs and live birds were shipped by train to city markets, producing $200,000 in income (equivalent to $2.8 million today).

In the 1950s, after the major boll weevil disaster in 1949, farmers looked again at poultry as a way to diversify. In 1950, C.S. Bridges and two other farmers created the Number Three Egg Producers Association near Patterson Springs. By 1957, the group had 76 members and was producing 1,000 cases (30 dozen to a case) of eggs each week. The Cleveland County Egg Producers Association, a similar group near Cherryville with 40 members, was producing 220 cases a week. Members of these groups often started small, with a few hundred hens, before increasing to a couple thousand birds. Eggs were delivered to a central location, where they were candled, washed, graded, and then shipped to grocery stores. Eggs from the Cherryville group were picked up each Wednesday and delivered to the warehouse for the A&P grocery store chain in Charlotte.

In 1922, R.E. Lawrence (right), Cleveland County farm agent and poultry enthusiast, demonstrates qualities to look for in a good bird. (Courtesy of University of North Carolina Libraries.)

Two young members of the Cleveland County Poultry Club man the booth at the state fair in Raleigh in 1929. The arch displays the traits needed for successful poultry production. The sign behind the boy reads "Produce Winter Eggs: one dozen eggs in December is worth three dozen in May." (Courtesy of North Carolina State University Libraries.)

Today, poultry standards are most often associated with show birds. But original breed standards were created first and foremost for production qualities. Breeds were classified as egg layers, meat birds, or dual purpose. Each breed had a standard for optimal production. Farm demonstrations like this one, about 1950, were used to teach breeders what qualities to select for good birds. (Courtesy of Earl Scruggs Center.)

Zeb Holland stands with his flock of Dominickers, a once popular breed of chickens around barnyards and homesteads. The dual-purpose breed was bred for both eggs and meat production. Hens were kept until they stopped laying. Most small flocks only had one lucky rooster—extra cockerels were destined for the frying pan. (Courtesy of Jim Holland.)

Carnie Elam and his son Jim examine their flock of chickens. At the time of this 1955 photograph, they were considered a large poultry operation, with a few thousand birds grown for egg and meat production. The feeder is made from a Gulf Multipurpose Gear Lubricant barrel. (Courtesy of North Carolina State University Libraries.)

In 1955, this poultry house at the Elam Poultry Farm near Fallston was considered state of the art for poultry production. (Courtesy of North Carolina State University Libraries.)

Jim Elam points to what appears to be either a fan used to keep eggs cool or some type of egg washer. Farmers had to deliver good-quality eggs to their distributor or cooperative. If eggs were too dirty, the Number Three Egg Producers Cooperative sent the eggs back to the farmer. After that happened a few times, farmers made sure to send clean eggs. (Courtesy of North Carolina State University Libraries.)

In 1972, the egg grading plant for the Number Three Egg Producers Cooperative, pictured in the background, built a small lagoon to hold the dirty wash water. The lagoon was built with the help of the Cleveland Soil Conservation District. Soil conservation technician Sam Jenkins examines the new lagoon. (Photograph by Jim Boggs, courtesy of Cleveland County USDA Service Center.)

This fair booth illustrates the progress of the Number Three Egg Producers Cooperative. The cooperative started with three members in September 1950 and had 65 members by 1954. (Courtesy of North Carolina State University Libraries.)

98

District conservationist Joe Craver (right) and Clemmie Royster examine a flock of white turkeys grazing a stand of fescue that has gone to seed. Grazing turkeys in good pasture like this cut down on feed costs and was common practice in the early to mid-1900s. (Courtesy of Cleveland County USDA Service Center.)

Turkeys have scratched and grazed this field down to bare ground. Notice that the chicken-wire fence is the only thing separating this flock of turkey from predators. (Courtesy of Cleveland County Cooperative Extension.)

Turkey production slowly shifted from ranged turkeys to housed turkeys. Below is a turkey house at John Hendrick's farm near Fallston in 1955. The turkeys were kept inside the house and had access to the top-story "sunporch." The bottom of the sunporch was made of wire to allow droppings to fall through, which were then scraped up and used as fertilizer. (Above, courtesy of Cleveland County Cooperative Extension; below, courtesy of North Carolina State University Libraries.)

Seven

From Horses to Horsepower

Horse and mules once powered Cleveland County farms. Farmers took pride in a good mule or workhorse, bragging about how many acres they could farm per animal. With mules and horses shipped in by train to sell, stables in Shelby were equivalent to modern-day car or tractor dealerships. In the early 1900s, among the stables in Shelby where farmers could buy and sell mules were Blanton's Stable and Doggett and Company's Stable. Good mules were always valuable, but the First World War put them at a premium. A 1917 advertisement for Blanton's Stable reads, "War Mules Wanted . . . good prices paid for mules from 5 to 10 years old, 900 to 1200 lbs."

After the war, the era of the mule and workhorse began to end. Some remember as children the day their fathers brought home the first tractor. For some, it was a special day, meaning the hours spent standing behind a mule and plow were over. In 1918, F.H. Lee bought the first Fordson tractor in Cleveland County from Chas L. Eskridge, who sold Ford cars and Fordson tractors in Shelby. By 1920, Eskridge had sold Fordson tractors to 39 farmers in Cleveland County. These tractors were steel-wheeled, advertised to take the place of eight mules but only costing as much as a team of two. Other early tractor dealers were W.A. Crowder and R.L. Hunt of Lattimore, who sold Cleveland tractors (based out of Cleveland, Ohio), Grady Mauney of Shelby, who sold Case kerosene tractors, and J.D. Lineberger of Shelby, who sold International tractors.

By the mid-1900s, more dealerships arrived. Allis-Chalmers, John Deere, International, Oliver, Minneapolis-Moline, Massey-Ferguson, Ford, and Case all had dealerships in the county. Farmers no longer had a favorite breed of draft horse or cross of mule. Instead, they had a favorite color of tractor. To be fair, in many instances, farmers were actually dependent on a single brand, as tractor implements at the time were not universal. A farmer who purchased an Allis-Chalmers tractor needed Allis-Chalmers implements. So farmers usually stayed true to a brand.

On Warren Street, the Doggett and Company Stable was the equivalent of a modern car or tractor dealership. In this early-1900s photograph, C.R. Doggett is holding the stick in the center in front of the stable hallway. Later, the Doggett family began selling cars and tractors in Shelby. (Courtesy of Earl Scruggs Center.)

A young colt grazes in a Cleveland County pasture. (Courtesy of Cleveland County Cooperative Extension.)

In the early 1900s, the sizes of farms were often described not by the number of acres they contained, but by the number of horses needed to work the farm. Real estate ads described two-horse farms, four-horse farms, six-horse farms, and more. A two-horse farm was the typical size farm in the county, 40 to 100 acres, while a six-horse farm likely contained around 300 acres. (Both, courtesy of Cleveland County Cooperative Extension.)

Although farms may have been described as two-horse farms, mules were likely doing the farmwork, as they were smarter and stronger than horses. At left, Zeb Holland continued to plow his garden with a mule into the 1960s, although he had a tractor. He believed the mule did a better job plowing. Below, a farmer cultipacks a field with a team of two mules before planting small grains. (Left, courtesy of Jim Holland; below, courtesy of Cleveland County USDA Service Center.)

In 1922, in front of a crowd in Lattimore, Forest Crowder is driving a Cletrac tractor, manufactured by the Cleveland Tractor Company, based out of Cleveland, Ohio. W.A. Crowder was a sale representative for the company and a prominent farmer in the county. For some of these folks, this may have been the first tractor they had seen in person. (Courtesy of Warren Crowder.)

Two Cleveland County farmers show off a Fordson tractor with a belly plow. Fordson was probably the most common steel-wheel tractor found in the county in the 1920s. For about a decade, starting in 1928, Ford stopped manufacturing tractors in the United States. Later, in 1938, Henry Ford and Harry Ferguson began manufacturing tractors again domestically with the production of the 9N. (Courtesy of Cleveland County Library.)

The caption on the back of this late-1920s photograph reads: "Beam Brothers, Shuford and Thamar of Waco, are growing enough legumes on their farm and which very few men in Cleveland County can boast of." (Courtesy of North Carolina State University Libraries.)

This appears to be the same Beam brothers and tractor from the previous photograph. The name on the plow, behind the young boy sitting in the furrow, is Oliver. Four companies merged in 1929 to form the Oliver Farm Equipment Company, which produced hunter-green-colored tractors and a full range of tractor implements. (Courtesy of Cleveland County Library.)

A tangled mess of vetch was no problem for this steel-wheeled tractor, but the ride must have been rough, judging from the cushion. A legume, vetch was commonly grown as a soil builder and plowed under before a cash crop. (Courtesy of Cleveland County Cooperative Extension.)

This farmer is operating a Farmall Cub with a sickle-bar mower. The Cub had 11 horsepower and was manufactured from 1947 to 1981, with few changes in design in that time span. (Courtesy of Cleveland County Cooperative Extension.)

The production of the Ford 8N started in 1947. The 8N had around 20 horsepower and was a very popular tractor. (Courtesy of Cleveland County Cooperative Extension.)

On September 9, 1972, Ab Hamrick used a Farmall C to prepare a strip of land for small grain planting. Strip-cropping was common in Cleveland County from the 1940s to 1970s. (Photograph by Jim Boggs, courtesy of Cleveland County USDA Service Center.)

A line of John Deere tractors, likely from C.J. Hamrick and Sons dealership in Boiling Springs, is on display at Miracle Farm Day at Gardner-Webb College. To the far left is the Model G, which was the largest tractor in the line. Miracle Farm Day was a massive undertaking that took months of planning. Thousands of people showed up, including Gov. W. Kerr Scott, to contribute labor

for the creation of a college farm at Gardner-Webb College. On August 31, 1950, seventy-five tractors, thirty-five bulldozers, and thousands of volunteers worked together under the direction of county farm agents and soil conservationists from Cleveland and surrounding counties. (Courtesy of Gardner-Webb University Archives.)

The Ivester family of Lawndale is trading in Allis-Chalmers tractors for newer models in 1953. Notice that the newer models came with mounted implements, an upgrade from the pull-type implements used with older models. (Courtesy of Ronnie Ivester.)

In 1957, David Dilling and Beverly Dorcas Plonk sit atop a Case tractor at the farm of John Bulter Plonk in Kings Mountain. This tractor likely came from the Case dealership in Kings Mountain, owned by Lewis Hovis. (Courtesy of Mary "Pucky" Nantz.)

Production began on the Ford Commander 6000 in 1965. At the time, this was considered a large tractor. Rated at 60 horsepower, the Ford Commander 6000 in this photograph is pulling a four-bottom plow with ease. The Hardin-Dixon Tractor Company on Washington Street sold Ford tractors. (Courtesy of Cleveland County Cooperative Extension.)

The Lutz-Yelton Tractor Company on Lafayette Street sold Farmall tractors—and a lot of them. Above, a farmer wearing dress shoes cultivates cotton on a Farmall M. Below, Bill Plonk of Kings Mountain sits on a Farmall C in 1966. (Both, courtesy of Cleveland County USDA Service Center.)

Eight

TERRACES, TRIBULATIONS, AND EARLY NO-TILL

Water is powerful. It can nourish or ruin a crop, farm, or even community. Early settlements in Cleveland County developed along streams not only because of fertile bottomland soil but because of power—waterpower. Waterwheels dotted the landscape, turning gristmills, sawmills, and early cotton gins. Some communities also developed around springs: McBrayer Springs, Sulpher Springs, Boiling Springs, and Lithia Springs, among others. Many of these communities developed a reputation for healing mineral water. In the late 1800s, hotels in Patterson Springs and Cleveland Springs were tourist attractions, drawing people who suffered from various ailments.

Water can be incredibly destructive. Heavy summer rains, rolling hills, and clay soils provided the perfect ingredients for erosion. After tillage and drying out, clay soils crust over, preventing water infiltration. Runoff from fields can pick up speed and power, dislodging particles of soil and carrying away fertile topsoil. During the late 1800s and early 1900s, with the arrival of widespread cotton farming, upland fields in Cleveland County lost on average between six and eight inches of topsoil. For Cleveland County farmers, finding ways to prevent and limit erosion was essential to long-term productivity.

Cleveland County farmers led the way in innovative soil conservation methods. According to the 1949 article "Terracing in the Southern Piedmont" by Arthur Hall, one of Cleveland County's first county farm agents, Dr. R.M. Gidney, was an enthusiastic advocate for terraces. Hall states, "By 1915 he had run level broad-base terraces on about 6,000 acres in Cleveland County. At that time Cleveland was one of the best-terraced counties in the state and was called 'the county of terraces.' " Cleveland County farmers also led the way with other erosion control methods. Many of these methods were promoted by the federal Soil Conservation Service and Cleveland Soil Conservation District. From cover crops to strip-cropping to no-till farming, the photographs in this chapter show the struggles and successes of farmers to control water and conserve soils.

This 1939 photograph is from the first field manual of the Cleveland Soil Conservation District and Soil Conservation Service. The manual is full of instructive photographs for constructing terraces. (Courtesy of Cleveland County USDA Service Center.)

After turning and loosening the soil on the high side, the farmer began the tedious process of earthmoving. This farmer uses a drag-pan, which looks similar to a wheelbarrow with no wheels, to scoop and move the soil from the high side to the low side. Other methods of constructing terraces relied solely on a turning plow, with rounds of soil being thrown both uphill and downhill to form a berm in the middle. For one terrace, a farmer might plow 50 rounds or more. Terrace construction was hard, time-consuming work. (Courtesy of Cleveland County USDA Service Center.)

A terrace is under heavy flood conditions at Coleman Self's farm in Shelby in 1962. The purpose of a terrace is to slow down and divert water, and this terrace is doing its job. Sometimes, terraces would overtop and burst. If a farmer did not notice and repair a broken terrace, a few large thunderstorms could quickly produce a gully. (Photograph by Samuel Jenkins, courtesy of Cleveland County USDA Service Center.)

This large gully at a Polkville farm started from a line ditch. Ditches were often used to mark property lines. Even kudzu was doing little to stop erosion. The gully was eroding into the field at a rate of 15 feet per year. To solve the problem, the farmer, with the help of the Soil Conservation Service, constructed a dam across the gully, creating a farm pond (Photograph by Samuel Jenkins, courtesy of USDA Service Center.)

Ditches alongside roads could also cause major gullies. To stabilize roadsides, the Soil Conservation Service and Cleveland Soil Conservation District promoted planting both kudzu and multiflora rose, providing farmers with seedlings of both. Both species grow incredibly fast and do a remarkable job holding down soil. Unfortunately, they grew so well that they are now invasive species. (Photograph by Samuel Jenkins, courtesy of Cleveland County USDA Service Center.)

In 1968, district conservationist Joe Craver proudly inspects a hedgerow of multiflora rose, planted to prevent erosion. (Photograph by Samuel Jenkins, courtesy of Cleveland County USDA Service Center.)

Overworking soils can do more harm than good. Reduced tillage not only helped maintain soil health but reduced fuel costs. One of the forerunning practices to no-till farming was mulch tillage. Mulch planting used coulters in front of a Cole-type corn planter to plant in roughly prepared land. This field at E.V. Allen's farm in Shelby was plowed only once on April 27, 1962, the date the photograph was taken. By traditional clean-cultivation standards, this field would have been considered far too rough and "trashy" for planting. (Photograph by Samuel Jenkins, courtesy of Cleveland County USDA Service Center.)

Here, a few weeks after the previous photograph was made, district conservationist Joe Craver examines milo seedlings coming up in the same field. Notice the amount of mulch and debris between the rows. (Photograph by Samuel Jenkins, courtesy of Cleveland County USDA Service Center.)

Mulch-planting led to strip tillage. This photograph shows a forward-thinking farmer, Harold Dellinger, experimenting with strip tillage in 1966. Attached directly to the three-point hitch is a ripper. Attached to the ripper is a ground-driven fertilizer hopper with a press wheel. Attached to the press wheel is a horse-drawn single-row planter. Modern sod planters vary little from this concept. Dellinger is planting a nine-year-old fescue field to corn. (Both, photograph by Samuel Jenkins, courtesy of Cleveland County USDA Service Center.)

As innovative as it was, this remained a two-man operation; someone still had to stand behind the planter. (Photograph by Samuel Jenkins, courtesy of Cleveland County USDA Service Center.)

This photograph, taken on May 18, 1966, shows Bill Plonk experimenting with strip tillage. He used two coiled shanks to prepare narrow, tilled strips in a recently grazed field. Also notice the fertilizer hoppers mounted to the middle of the tractor. (Photograph by Samuel Jenkins, courtesy of Cleveland County USDA Service Center.)

After preparing the strips, Plonk makes one more a pass to plant the corn. (Photograph by Samuel Jenkins, courtesy of Cleveland County USDA Service Center.)

The results of Plonk's strip-tillage experiment were promising. Here, he has a good stand of corn. (Photograph by Samuel Jenkins, courtesy of Cleveland County USDA Service Center.)

This 1972 photograph shows an early Allis-Chalmers planting rig adapted for no till. Danny Cochran is planting barley for grain production directly into a field with no land preparation and heavy corn residue. He is spraying and planting in the same pass, drastically reducing fuel costs for tillage. Since corn residue remains on top of the surface, erosion is nearly eliminated. Cochran actually made another pass to plant the row middles, since his planters were spaced for corn and cotton instead of small grains. (Photograph by Jim Boggs, courtesy of Cleveland County USDA Service Center.)

Danny Cochran (left) and soil conservationist Sam Jenkins are examining the barley growth. (Photograph by Jim Boggs, courtesy of Cleveland County USDA Service Center.)

Corn is being planted no-till into small grain residue at F.S. Dedmon's farm on May 31, 1972. Notice that the combine in the background is harvesting grain. The field was planted in corn on the same day it was harvested for grain. (Photograph by Jim Boggs, courtesy of Cleveland County USDA Service Center.)

This field contains no-till cotton planted directly in wheat stubble with no land preparation. Farmer Roy Cochran grins in approval. (Photograph by Jim Boggs, courtesy of USDA Service Center.)

BIBLIOGRAPHY

Daniels, R.B. "Soil Erosion and Degradation in the Southern Piedmont of the USA." *Land Transformation in Agriculture*. New York: Wiley, 1987.

Davis. J.R. "Reconstruction in Cleveland County." *Historical Papers*. Durham, NC: Trinity Press, 1914.

Farm Maps Directory and Business Guide: Cleveland County, North Carolina. Rockford, IL: Rockford Map Company, 1954.

Hall, Arthur R. "Terracing in the Southern Piedmont" *Agricultural History* (April 1949): 96–109.

Kendrick, Larkin S. *Larkin S. Kendrick Papers: 1861–1874*. State Archives of North Carolina.

Olmsted, Frederick Law. *A Journey in the Back Country*. New York: Mason Brothers, 1860.

Vanatta, E.S., and F.N. McDowell. *Soil Survey of Cleveland County, North Carolina*. Washington: US Department of Agriculture, 1918.

Weathers, Lee Beam. *The Living Past of Cleveland County*. Shelby, NC: Star Publishing Company, 1956.

ABOUT THE ORGANIZATION

Conservation districts were created in the 1930s in response to the Dust Bowl, a decade-long wind erosion event centered in the Great Plains. The federal government created the Soil Conservation Service, now known as the Natural Resources Conservation Service (NRCS). It also passed legislation allowing states to create soil conservation districts. Local farmers and landowners gathered and elected a local board of supervisors to oversee conservation efforts. The purpose of the Soil Conservation Service and soil conservation districts was to educate farmers and landowners on the importance of preventing erosion. They also provided advice and financial incentives to farmers interested in protecting soil.

The first conservation district in Cleveland County was the Broad River Soil Conservation District. Formed in 1938, it encompassed Cleveland, Rutherford, and Polk Counties. Like in the Great Plains states, local soil was highly eroded; however, rain erosion was the culprit in North Carolina. Researchers believe the Piedmont and Foothills lost on average six inches of topsoil between the late 1800s and early 1900s with the arrival of cotton farming. Eventually, in the western Piedmont, the sight of red clay and gullies would become common, as much of the topsoil had been washed away.

The Broad River Soil Conservation District promoted contour planting, terracing, strip cropping, and many other conservation practices. In the early years, the district provided farmers with labor from the Civilian Conservation Corps to implement practices and had a tractor that farmers could rent to build terraces. The district provided farmers with kudzu and white pine seedlings to stabilize eroding areas.

In 1961, the Broad River Soil Conservation District split into three districts corresponding with county boundaries. The Cleveland County Soil and Water Conservation District continues to work with farmers and landowners to implement good land stewardship.

The following Cleveland County farmers have served as supervisors for the Broad River Soil Conservation District or Cleveland Soil and Water Conservation District: Yates Brooks, Tom Cornwell, Wayne Ware, Cameron Ware, Ralph Spangler, Harold Plaster, Dan Jones, Samuel Jenkins, Roy Dedmon, Robert Sweezy, Paul Davis, Carl Debrew, Bill Walker, Jim Boggs, Bill Plonk, Randy McDaniel, Roger Eaker, Jeff Cornwell, Ted Wortman, Michael Underwood, and Sherri Greene.

Visit us at
arcadiapublishing.com

www.ingramcontent.com/pod-product-compliance
Lightning Source LLC
Chambersburg PA
CBHW080613110426
42813CB00006B/1494